# THE TWIN PARADOX

## A RELATIVELY POSSIBLE STORY

WRITTEN AND ILLUSTRATED
BY JOHN SIEGLAFF

**With special thanks**, first and foremost, of course, to my mother whose faith keeps me going; to my sister, Kimberly, whose help guided me through many a minute detail as well as the grand scheme of this story; to my Aunt Barb for her amazing support, generosity, and artist's eye; to Katie for her sharp critic's eye and helpful suggestions throughout; to my brother, Mark, whose publishing knowledge made this book possible; to my high school physics teacher, Mr. Koehler, for delivering the concepts of physics in the most captivating ways imaginable; to my Aunt Mary Ann for her invaluable advice; and to all my friends who showed their interest and support throughout the duration of this project. I couldn't have done it without you.

Copyright © 2017 by John Sieglaff
All Rights Reserved.
Library of Congress Catalogue Card Number: 2017913192
ISBN: 978-0-692-93023-6

Published by John Sieglaff

Library of Congress Cataloging-in-Publication Data

Sieglaff, John J., 1987-
The Twin Paradox Summary: When Chase and Eddie — ten-year-old identical twins — are assigned a science fair project, the results end up changing their lives forever / John J. Sieglaff; illustrated by John J. Sieglaff.–1st ed.
p.cm
ISBN 978-0-692-93023-6
Physics – United States – Childrens Literature – I. Sieglaff, John J., 1987-
2017913192

Written and Illustrated by John Sieglaff
Printed in the United States
First Printing: 2017

**For my grandfather**
whose questions of
the cosmos piqued my
curiosity when I was
only ten

Chase and Eddie were only ten—
Identical twins and carefree when...

Their teacher assigned a science fair project
To build and test a functional object.

Straightaway Eddie knew just what he'd do:
Build his own drumset, then bang on it too!

But while Chase went straight to work as well,
Just what it was, he would not tell;

No guesses at it could be said,
He kept it hidden away in the shed.

But time soon came to test his secret
And so he let his brother peek it...

Eddie's eyes bulged at Chase's machine
It was unlike anything he'd ever seen!

A Radio Flyer with buttons galore,
Em Drive Thruster and so much more.
Rusty, makeshift, and quite unsturdy...

Was this thing s'posed to be 'SPACEWORTHY'?!
"Well," said Chase. "Is she the best ship you've seen?"

"You're not really going to FLY that thing?!
You'll hit the ground before you reach the air!
It's sure to lose at the science fair."

But Chase knew better than let his brother impede
"Just wait until I tell you about its super-top speed!
She may not look like much at first sight,
But this baby cruises at the speed of light!"

"Go ahead, laugh and poke fun at me
While I race laps through the galaxy!"

And with one month till the project was due,
Chase thought he'd better start his mission soon!

He wasted no time the very next day,
Starting his morning by flying away.

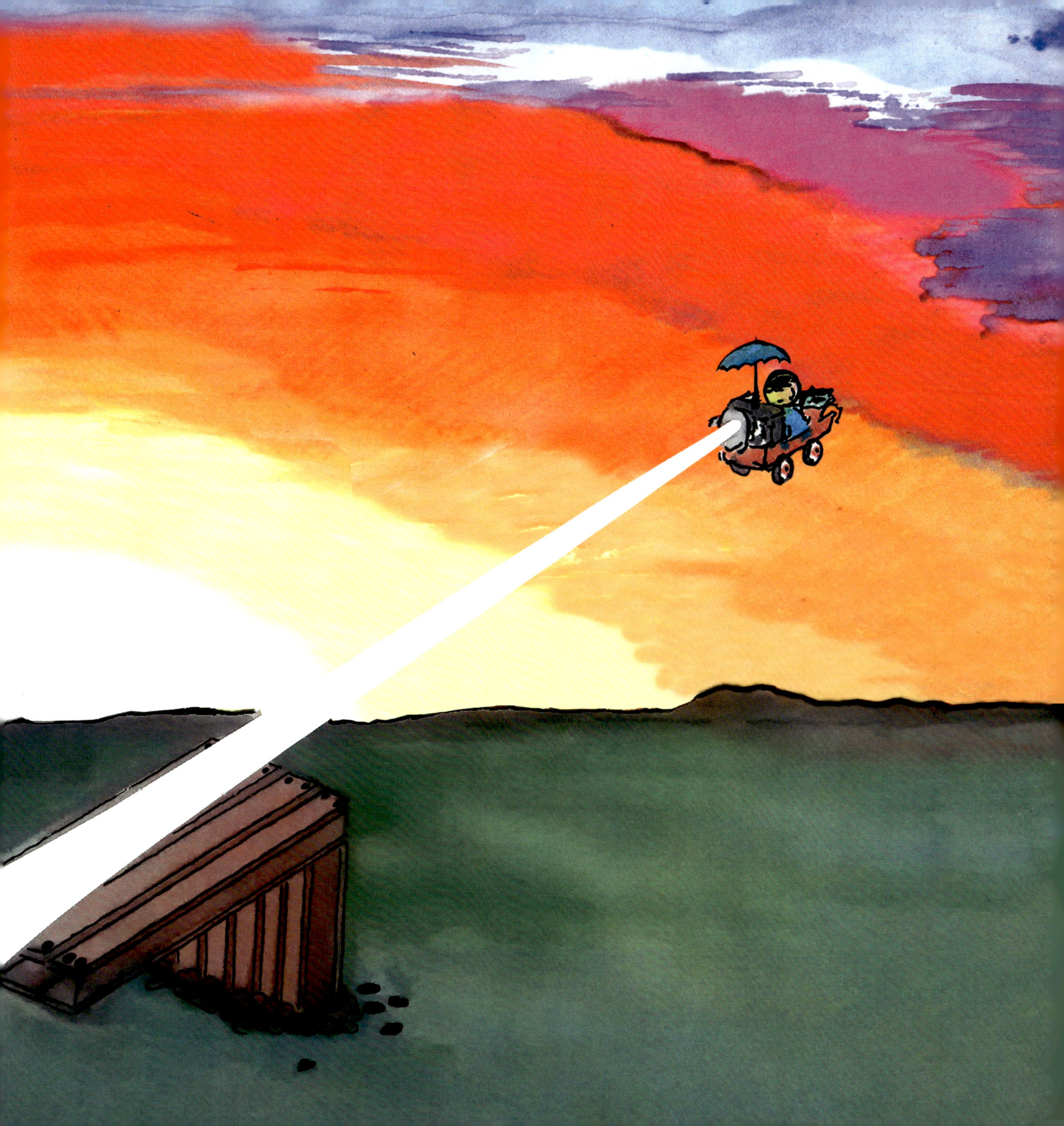

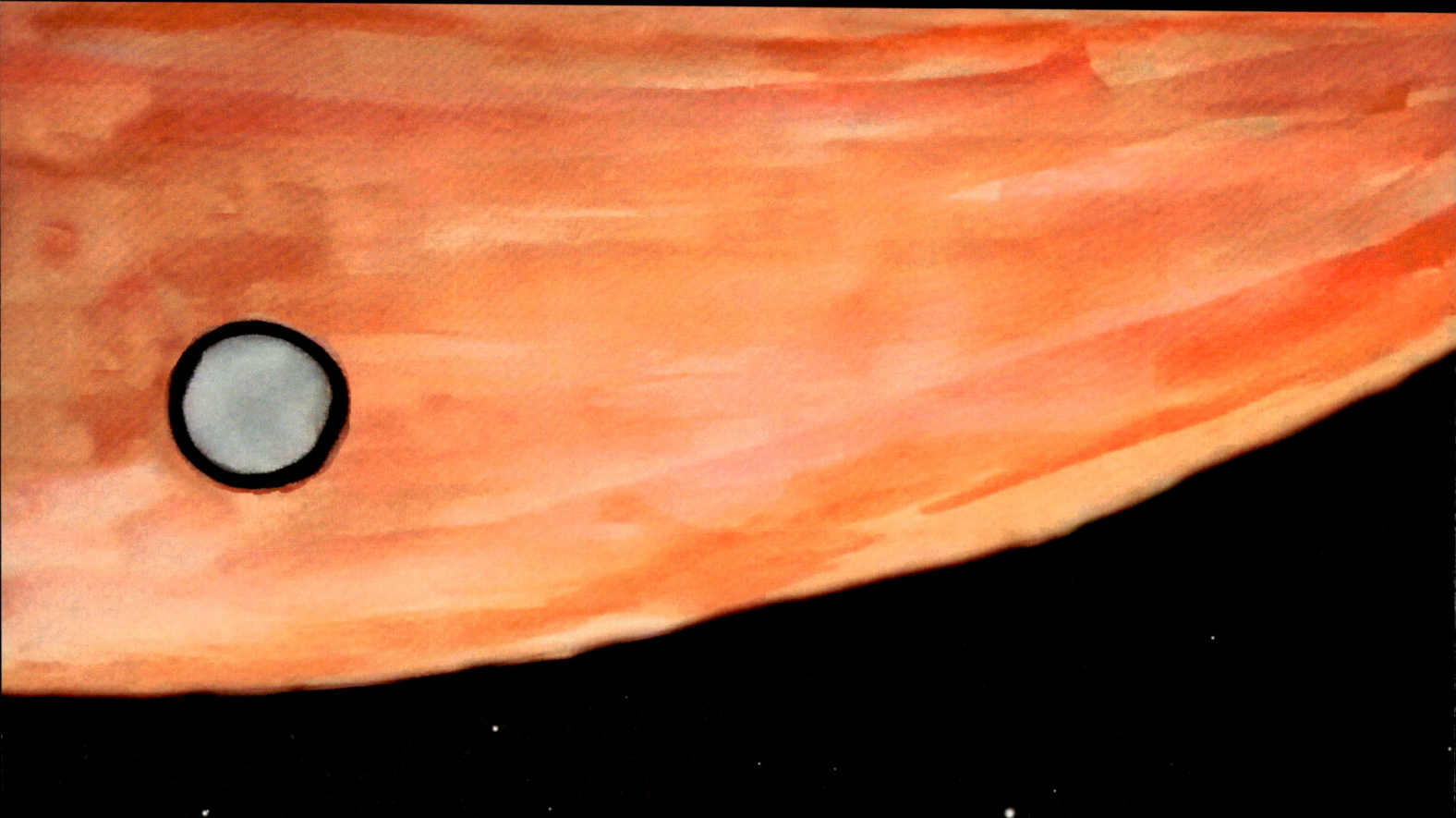

It didn't take long till Mars was passed,
Reaching incredible speeds—and FAST!!

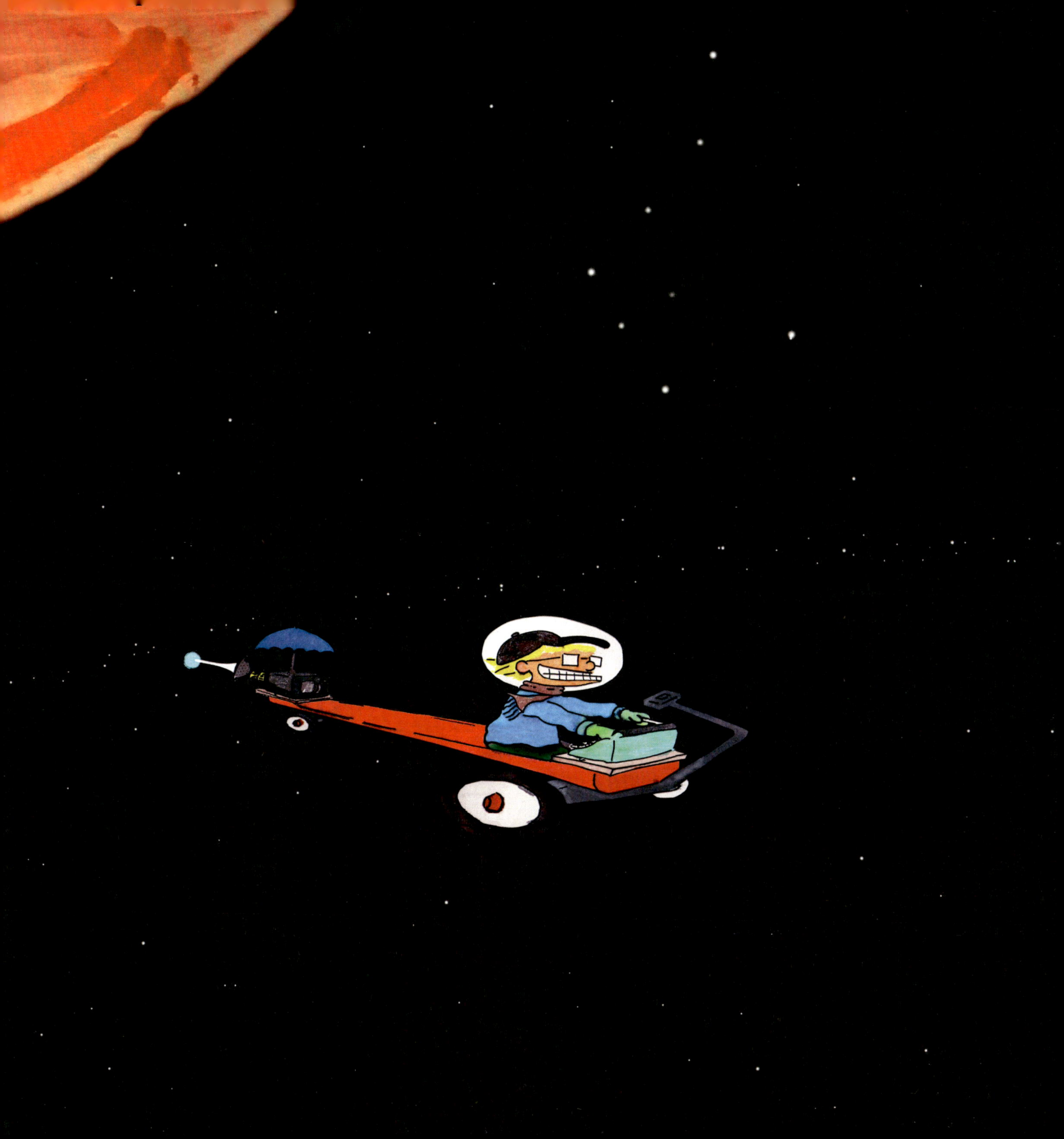

Chase whooped and hollered and shouted out, "YES!!"
For his unlikely spaceship was quite the success!

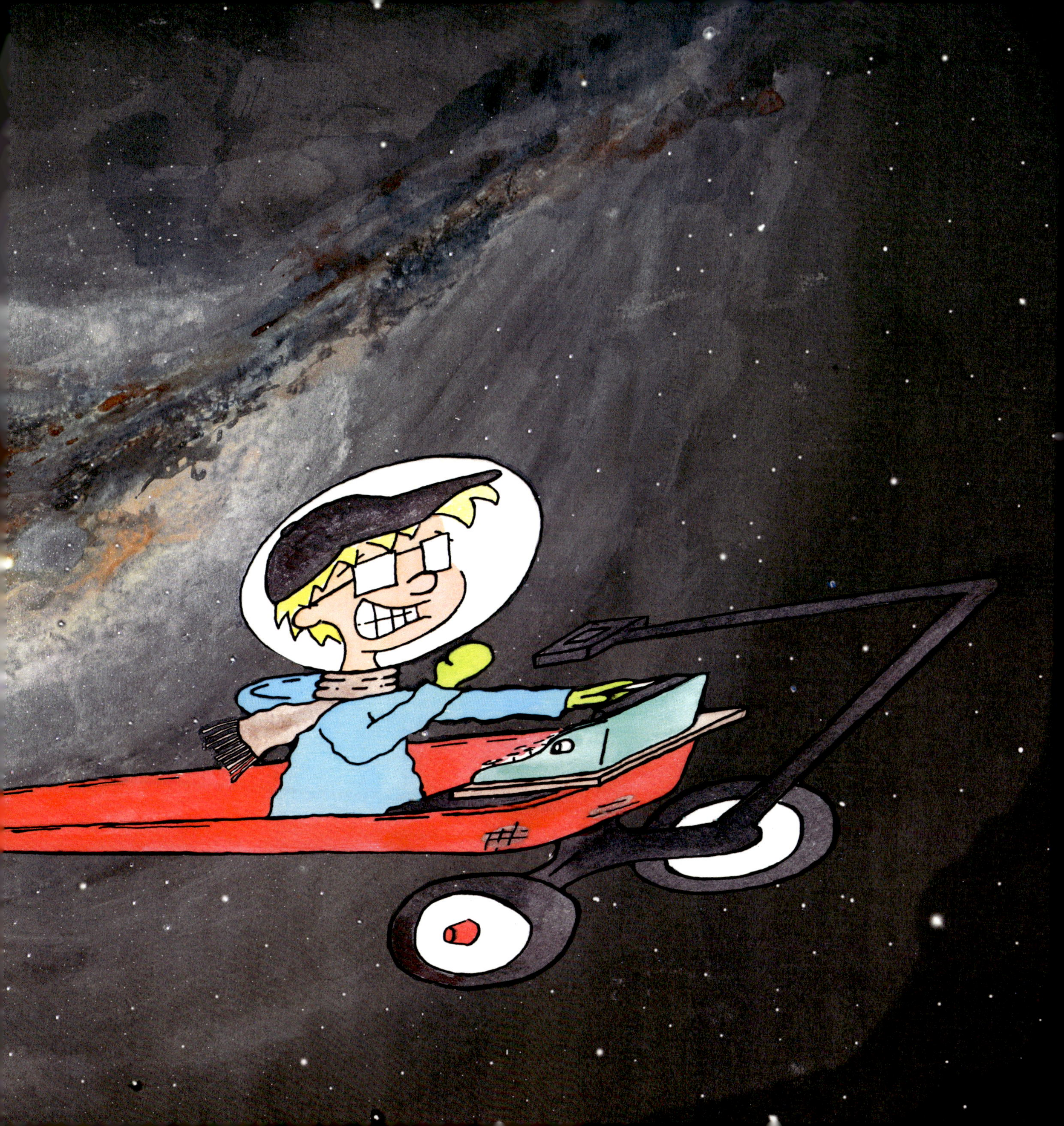

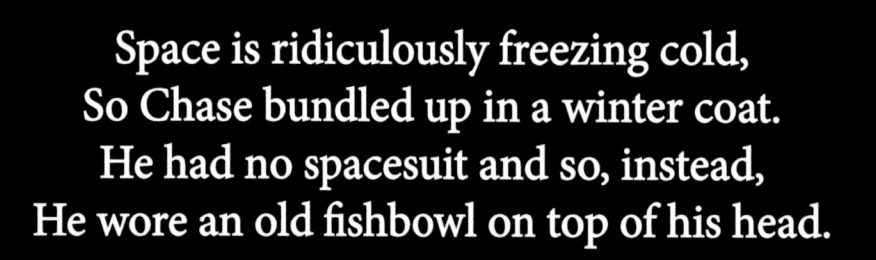

Space is ridiculously freezing cold,
So Chase bundled up in a winter coat.
He had no spacesuit and so, instead,
He wore an old fishbowl on top of his head.

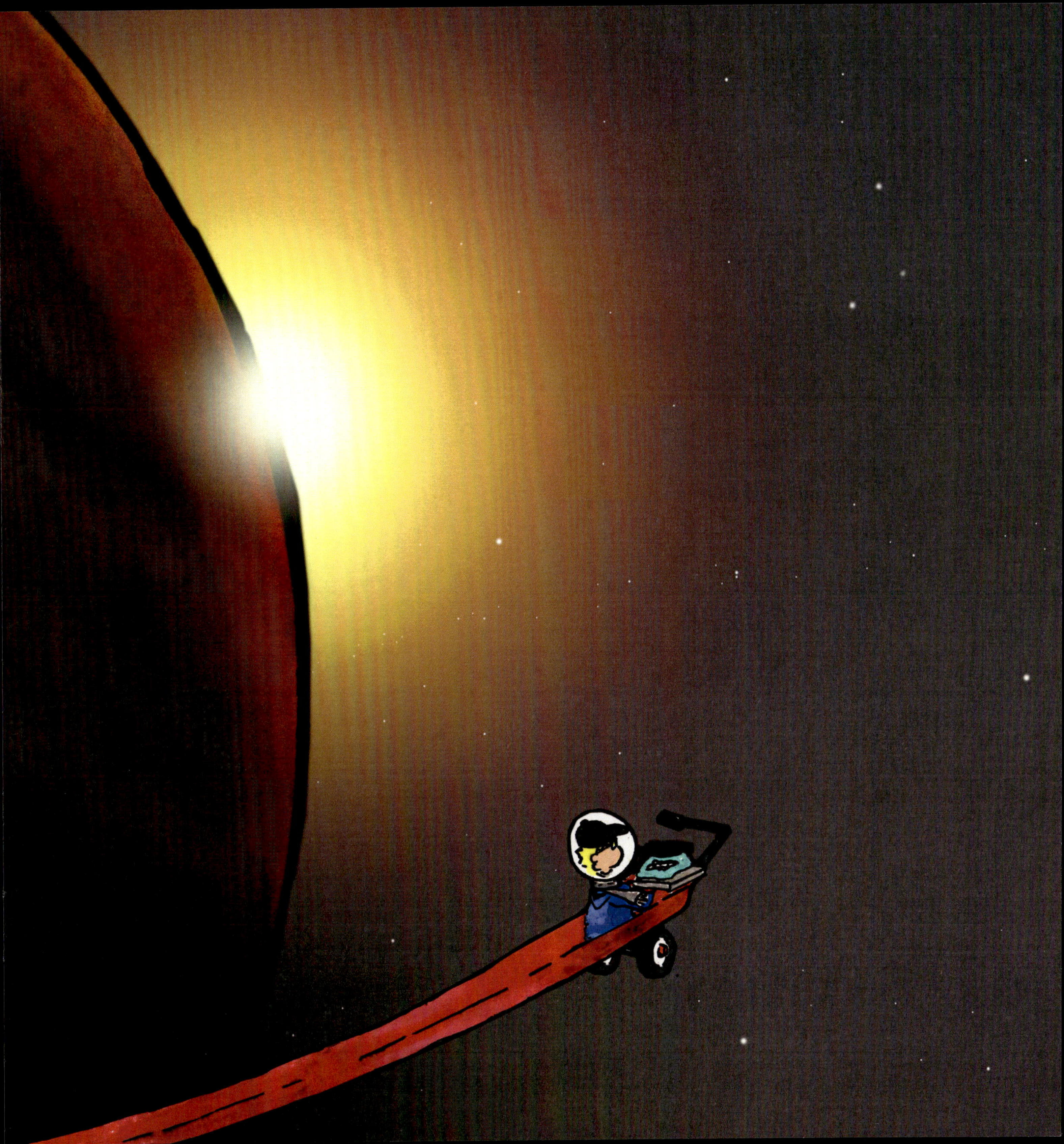

He brought comics and snacks along for the ride
But spent most of his time just staring outside.
The world whizzing by is a breathtaking sight
When zipping through space at the speed of light.

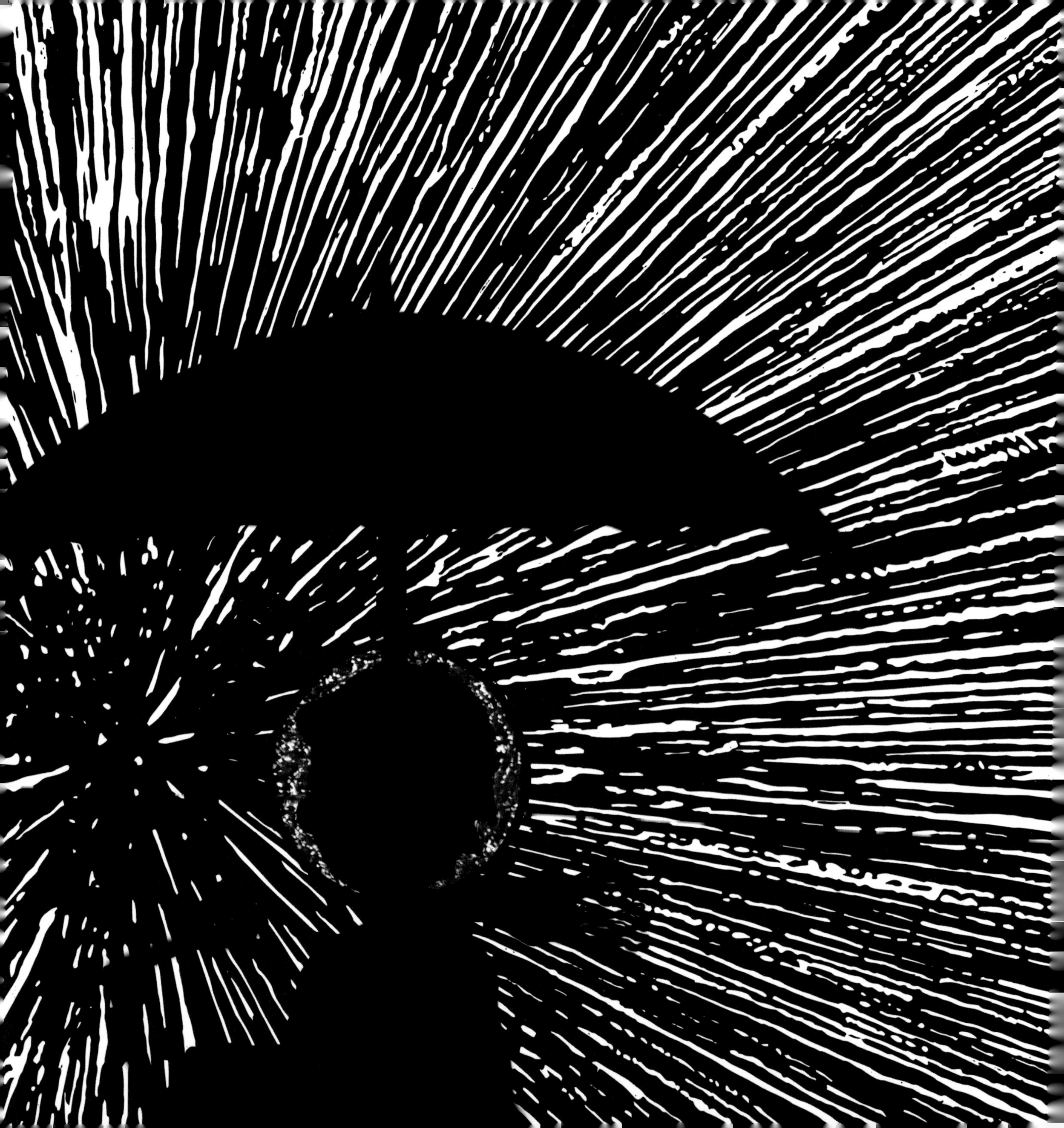

Chase hoped his family wouldn't miss him too much,
He explained in a note he'd be back in a month.

But when a month passed and Chase hadn't returned,
Indeed his folks grew upset and concerned.

Chase cruised around space for thirty days—HIS time...

Meanwhile, Eddie finished school, found love, built a home, got married...

...had a kid, wrote a book, biked the country, walked his daughter down the aisle, became a grandfather...

And when he turned sixty-nine...

A blast in the sky caught everyone's eye.

And a flash from the crash turned Eddie outside.

Chase back from space?! Could it really be true?!
Still in his cap, not a day older too...

Not a day older—WOW! Now, how could that be?
Eddie exclaimed, "Somebody pinch me!"

The two stood and stared as confused as could be...
Wondered, and pondered, and mused fruitlessly.

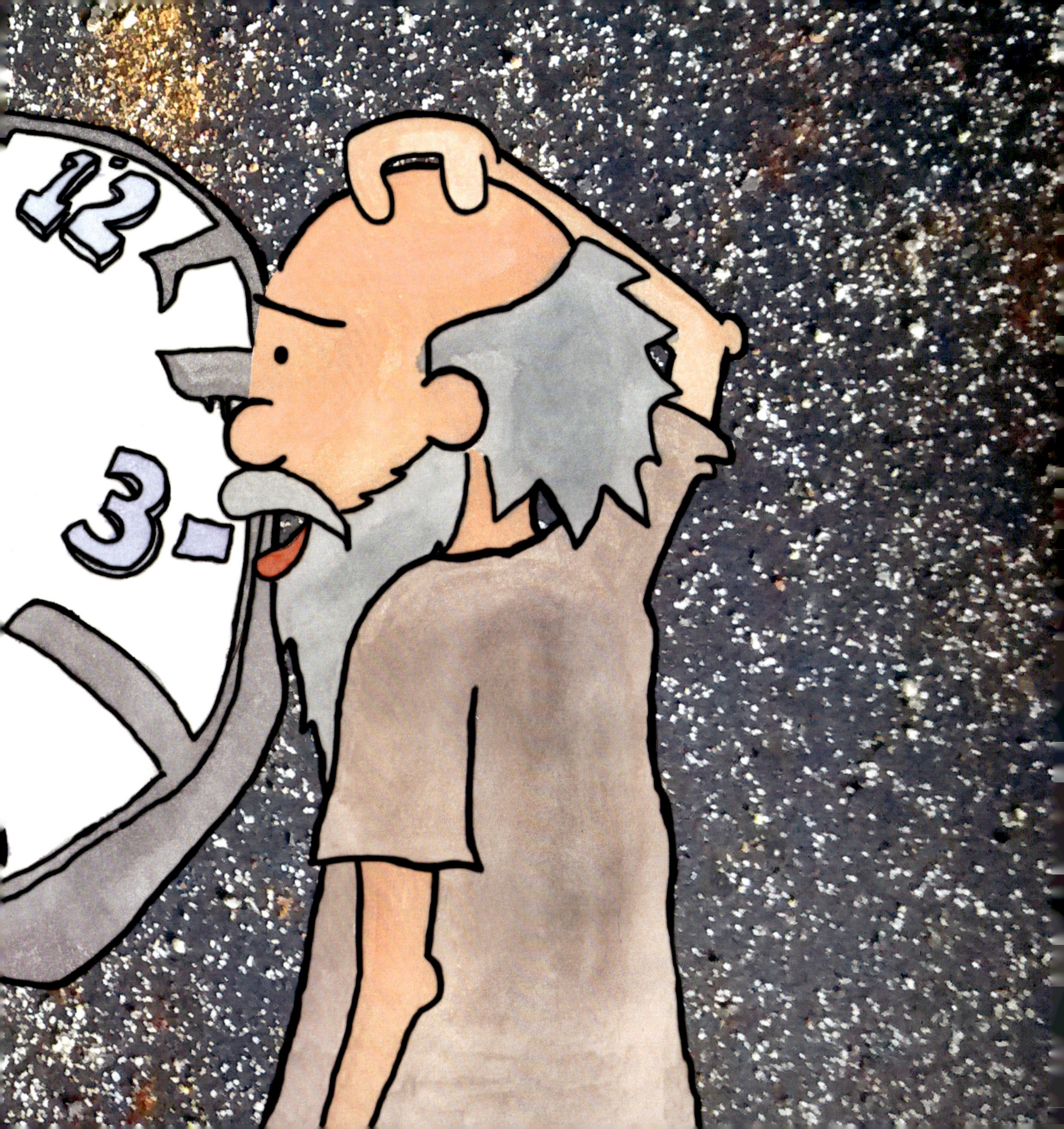

The Twin Paradox, no one could relate
To this weird phenomenon, strangest of fates.
Chase managed to get along alright,
Despite the results of moving like light.

But as Eddie once so wisely noted...

"Age misalignment's not quite the object,
The point is, you're SO late on your science fair project!"

# THE

END

# What 'A Relatively Possible Story' Means:

'The Twin Paradox' goes back to 1905 when Albert Einstein first formulated his Theory of Special Relativity. To help others (and himself) understand it better, he imagined what would happen to twins if one of them traveled at speeds close to the speed of light — the fastest speed in our universe.

There are a number of moments in this illustrated telling of Einstein's thought exercise that are not based in reality; you certainly can't fly to outer space in a Radio Flyer — the same way a bulky coat and fishbowl are no replacement for a spacesuit (don't try this at home, kids)! The idea, however, that time wouldn't move at the same rate for the twins is, indeed, accurate!

See, space and time are actually just two parts of a single thing called **spacetime** and the more you move through one part, the less you move through the other. Because Chase traveled at such a high speed, he moved through more space than he did time and, therefore, didn't age at the same rate that Eddie did. We can travel through time at different rates based on how fast we move through space! It sounds crazy, but it's true!

Physics is full of seemingly impossible truths about the nature of our world. I hope The Twin Paradox has inspired you to seek out more information about this most-intriguing study on the science of our universe!